SAXON MATH™

Intermediate 3

Learning Stations

A Harcourt Achieve Imprint

www.SaxonPublishers.com
1-800-284-7019

ISBN-13: 978-1-6027-7415-5

ISBN-10: 1-6027-7415-3

© 2008 Harcourt Achieve Inc.

All rights reserved. No part of the material protected by this copyright may be reproduced or utilized in any form or by any means, in whole or in part, without permission in writing from the copyright owner. Requests for permission should be mailed to: Paralegal Department, 6277 Sea Harbor Drive, Orlando, FL 32887.

Saxon is a trademark of Harcourt Achieve Inc.

Printed in the United States of America

3 4 5 6 7 1186 19 18 17 16
4500622033

Learning Stations

Learning Station activities provide opportunities for students to reinforce and extend their learning in interactive and cooperative settings.

Use of the Learning Station activities depends on the goals for instruction, the needs of the students, and the specific learning activity. The activities may be assigned to:

- build student background prior to a new increment of instruction
- reinforce and extend a new increment of instruction
- reteach a skill or concept following instruction and assessment

The activities are to be completed by students working in small groups or individually.

Assigning Groups

Group members may be assigned randomly or based on various criteria, such as grouping students who need help with a specific skill or assigning students to play specific roles within a group (such as recorder or reporter).

Types of Groups

Students may be paired or placed in larger groups. For pairing, students may be assigned partners on a weekly or monthly basis. Pairing activities are easy to manage on a daily basis and give students one-on-one interaction as they practice skills. Placing students in groups of three or more provides cooperative learning experiences.

Daily Reinforcement

A reteaching blackline master is provided to support each lesson. The reteaching master activities reinforce the skills or concepts taught. They may also be used to assess the effectiveness of the Learning Station activities.

Materials List

The Materials List identifies items needed to assemble the Learning Stations for each lesson.

Lesson	Activity	Materials	Reteaching
1	Divide students into groups of four. Write three similar date questions on the board. • Each group answers the first question together and compares answers. • Divide each group into pairs to answer the second question. • Each member solves the third question on his or her own and then checks his or her answer with the group.	• calendar from Lesson 1 (or classroom or other calendar)	1
2	Teach students having difficulty finding the rule this visual strategy. • Write the following counting pattern on the board: 3, 6, 9, 12,... • Have students copy the pattern on their paper. • Draw a line up from the 3 and down to the 6. Record the change above the numbers. +3 • Have students record the change for each consecutive term in the sequence. • Have students identify the rule. count up 3 • Have students apply the rule to find the next three terms. 15, 18, 21 +3 +3 +3 +3 3, 6, 9, 12, ____, ____, ____,	• pencil and paper	2
3	Some students may have difficulty remembering to skip count by 5 when reading minutes on a clock. Help them make this visual reminder of the counting pattern. • Direct students to draw a curved arrow outside the clock face from 12 to 1. • Above the arrow, students should write "5." • Next, students draw another arrow from 1 to 2 and write "10" above it. • Students should continue drawing arrows from each number to the next number until they get back to 12. • They should continue by writing the "skip counting by 5" numbers above each arrow. • Students can keep this page in their math folder as a reminder of this counting pattern. 60 5 10 15 20 25 30 35 40 45 50 55 12 1 2 3 4 5 6 7 8 9 10 11	• extra copies of Lesson Activity 2	3

Copyright © by Harcourt Achieve. All rights reserved.

Lesson	Activity	Materials	Reteaching
4	Ask students which numbers the letters might represent and why the letters might represent those numbers. 	• number line with three points labeled with letters	4
5	Provide students with a copy of a large circle divided into quarters. ***Say, "Each of the 4 pieces of our circle is called a quarter. Four quarters make a whole."*** • Next ask the students to pretend that the circle is a clock. Where the 12 would be, students should write ":00". Where the 3 would be, they should write ":15". Where the 6 would be, they should write ":30". Where the 9 would be, they should write ":45". ***Say, "This picture can help you remember half and quarter hours and how to write the time in digital form when the minute hand points to these numbers on the clock."***	• copy of large circle divided into quarters, one per student	5
6	Students who have difficulty with basic addition facts can use a number line to better understand this concept. • Write "5 + 7" on the board. • Remind students that when we add, the numbers get bigger, so we move up or to the right on the number line. • Tell students that we start at 5 and jump up the number line 7 times to find the sum. • Model finding the sum on the number line by starting at 5 and jumping 7 times to 12. • Write "8 + 6" on the board. • Direct students to put their finger or the eraser end of their pencil on the 8. • Remind them to jump up the number line 6 times. ***Say, "What is the sum?"*** 14 • Repeat with different addition facts. 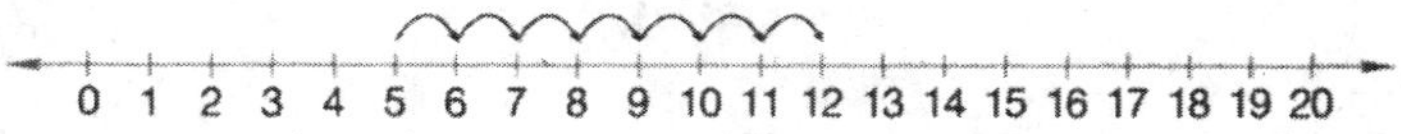	• number line 0–20, 1 per student	6

Copyright © by Harcourt Achieve. All rights reserved.

Lesson	Activity	Materials	Reteaching
7	Students who have difficulty with basic subtraction facts can use a number line. • Write "15 – 7" on the board. • Remind students that when we subtract, the numbers get smaller, so we move down or to the left on the number line. • Tell students that we start at 15 and hop down the number line 7 times. • Model finding the difference on the number line by starting at 15 and hopping 7 times to 8. • Write "13 – 3" on the board. • Direct students to put their finger or the eraser end of their pencil on the 13. • Remind them that they will hop down the number line 3 times. ***Say, "What is the difference?"*** 10 • Repeat with different subtraction facts.	• number line 0–20, 1 per student	7
8	Some students may have difficulty visualizing the relationships between fact-family members. Use manipulatives to make the relationships more concrete. • Write "+ + – –" in a column on the board. Write the facts in the column as students model them with manipulatives. • Direct students to show 5 + 3 with groups of 5 tiles on the left and 3 tiles on the right. • Switch the position of the groups. ***Say, "Does the sum change?"*** no • Combine the groups. Show 8 – 3 by removing 3 tiles from the group. ***Say, "What is the difference?"*** 5 • Combine the groups again. Show 8 – 5 by removing 5 tiles from the group. ***Say, "What is the difference?"*** 3	• color tiles, 20 per student	8
9	Students who have difficulty finding missing addends can use a number line. • Write this problem on the board: 3 + □ = 7. • Remind students that since we do not know how many hops to make, we will start on 3 and count the hops until we get to 7. • Model finding the missing addend on the number line by starting at 3 and counting the hops to 7. • Write this problem on the board: 7 + □ = 13. • Direct students to put their finger or the eraser end of their pencil on the 7. • Remind them to count the hops until they get to 13. ***Say, "What is the missing addend?"*** 6 • Repeat with different missing-addend facts. 1 2 3 4 0 1 2 3 4 5 6 7 8 9 10 11 12 13 14 15 16 17 18 19 20	• number line 0–20, 1 per student	9

Copyright © by Harcourt Achieve. All rights reserved.

Lesson	Activity	Materials	Reteaching
10	Have students take out the pennies from their money manipulatives. ***Say, "Use your pennies to model the following number sentence I have written on the board: 4 + 5 + 7 = □. Combine two groups of pennies. It doesn't matter which two groups you combine. How many pennies are in your group?"*** 4 + 5 = 9, 5 + 7 = 12, 7 + 4 = 11 • Write student replies on the board. ***Say, "Now, I want you to combine the last group of pennies with the other group. How many pennies do you have in the group now?"*** 16 ***Say, "We had different answers after adding the first two numbers together, but when we finished adding, we all had the same answer."*** • Continue this activity if desired using other problems such as 2 + 8 + 5, 7 + 1 + 6, and 4 + 4 + 9.	• pennies from money manipulatives	10
11	Some students may have difficulty with the abstract concept of place value. Base ten blocks on a place-value chart will help them visualize the concept. • Review the value of each block as you place it on the appropriate section of the place-value chart. ***Say, "A unit cube has a value of one. Ten ones are equal to one ten. Ten tens are equal to one hundred."*** • Direct students to place 3 ones, 2 tens, and 1 hundred on their place-value chart. ***Ask, "What number are you showing?"*** 123 ***Say, "Add 3 more tens. What is the number now?"*** 153 ***Say, "Show 256 on your place-value chart."*** 2 hundreds, 5 tens, 6 ones ***Ask, "How many hundreds?"*** 2	• base ten blocks • place-value chart (or Lesson Activity 6) for each pair of students	11
12	Write three-digit numbers in written form on chart paper and then have students make three-digit numbers with digit cards.	• digit cards 0–9 • chart paper	12
13	Have students individually solve the example without looking at the solution. Then divide the students into groups of four. • Students take turns telling the group the strategy they used to solve the problem. All strategies are acceptable. • One person from the group should be chosen to be the recorder. The group works together to write a summary of their strategies using words, numbers, or both.	• money manipulatives • counters	13
14	Have students write and solve their own two-digit subtraction problems and check them by modeling them with base ten blocks.	• base ten blocks • pencil and paper	14

Copyright © by Harcourt Achieve. All rights reserved.

<table>
<tr><th>Lesson</th><th>Activity</th><th>Materials</th><th>Reteaching</th></tr>
<tr><td>15</td><td>Some students may need a visual representation to round to the nearest 10. Teach them this strategy using a hundred-number chart.
• Write “78” on the board.
• Tell students to find the column where all numbers end in 5. They should place their pencils on that column.
• Point out to students that if the number they are rounding comes before the pencil, they round down. If the number is on the pencil column or comes after the pencil, they round up.
• The number 78 comes after the pencil so it rounds up to 80.</td><td>• hundred-number chart
• pencil</td><td>15</td></tr>
<tr><td>16</td><td>Some students may have difficulty with regrouping. Organizing their money manipulatives on a place-value chart will help them see the connection between place value and regrouping.
• Write “$487 + $336” on the board in vertical form.
• Students model each number using money manipulatives in the appropriate space on their place-value chart.
• Direct students to combine the $1 bills.
Ask, “What is the total?” 13
Ask, “What do we need to do?” Regroup 10 ones as 1 ten.
• Direct students to combine the $10 bills.
Ask, “What is the total?” 12
Ask, “What do we need to do?” Regroup 10 tens as 1 hundred.
• Record the sums and trades on the board as students model them with money manipulatives.
Ask, “How many hundreds?” 8
Ask, “What is the sum?” $823</td><td>• place-value chart
• money manipulatives</td><td>16</td></tr>
<tr><td>17</td><td>If students need extra help with comparing numbers, provide them with this modified place-value chart:
<table><tr><th></th><th>Hundreds</th><th>Tens</th><th>Ones</th></tr><tr><td>1st number</td><td>2</td><td>6</td><td>3</td></tr><tr><td>2nd number</td><td>2</td><td>7</td><td>1</td></tr><tr><td>Comparison</td><td>=</td><td><</td><td></td></tr></table>• Write “263 _?_ 271” on the board.
• Students write the numbers in their place-value chart.
• Compare the digits in the hundreds place. Since those numbers are the same, write “=” in the comparison row.
• Compare the digits in the tens place. Since the first number, 6, is less than the second number, 7, write “<” in the tens place in the comparison row.
• The overall comparison is determined by the first symbol that is not an equal sign as we read from left to right.
263 < 271</td><td>• modified place-value chart</td><td>17</td></tr>
</table>

Copyright © by Harcourt Achieve. All rights reserved.

Lesson	Activity	Materials	Reteaching
18	If students require additional help with recognizing and working with "some and some more" stories, provide this visual reminder of the pattern. • Write the following template on the board: ________ + ________ = ________ some some more total • Students copy the template on their card. Students should keep this card in a handy place and use it when they do Lesson Practice and Written Practice problems.	• notecards or index cards	18
19	Explain that rent is payment made to use someone's land or property. ***Say, "You might know someone who pays rent every month. What kinds of places can you rent?"*** house, apartment, boat, office, condo, warehouse, storage facility ***Say, "Some board games use property and rent. Every time Anna lands on Michel's property, she has to pay $100. Anna landed on Michel's property 3 times. How much rent did Anna pay Michel altogether?"*** $100 + $100 + $100 = $300 • Ask students to draw a house, apartment, or property for a board game. Then have partners work together to decide how much each place would cost to rent.	• pencil and paper	19
20	If students require additional help with recognizing and working with "some went away" stories, write this visual reminder of the pattern. Write the following template on the board: ________ − ________ = ________ some − some went away = what is left • Students copy the template on their card. • Students should keep this card in a handy place. They can use it when they do Lesson Practice and Written Practice problems.	• blank notecard for each student	20
21	Have students form pairs to practice adding dollars and cents. • Write an addition story problem involving money on the board. Ask students to think about how they would solve the problem. • Students work in pairs to write the answer. Volunteers share their answers with the class.	• pencil and paper	21
22	If students have difficulty aligning the digits in money amounts, provide grid paper for them to use. • Show students an example of a money addition problem aligned on grid paper. Allow them to use grid paper whenever needed to align problems. Another option is for students to turn their regular lined paper sideways so that the lines on the paper are vertical. $ 5 . 2 7 + $ 2 . 3 4 $	• grid paper	22

Copyright © by Harcourt Achieve. All rights reserved.

Lesson	Activity	Materials	Reteaching
23	Some students may have difficulty with regrouping in subtraction. Organizing their money manipulatives on a place-value chart will help them see the connection between place value and regrouping. • Write "\$374 – \$156" on the board. ***Ask, "How much do we start with?"*** \$374 • Students model \$374 using money manipulatives in the appropriate spaces on their place-value chart. ***Ask, "How much are we taking away?"*** \$156 ***Ask, "Where do we start?"*** ones place ***Ask, "How many ones are we taking away?"*** 6 ***Ask, "Do we have enough ones?"*** no ***Ask, "How do we get more ones?"*** We trade 1 ten for 10 ones. • Students trade one \$10 bill from the tens place for 10 \$1 bills which they put in the ones place on the place-value chart. ***Ask, "How many ones do we have now?"*** 14 ***Ask, "How many tens?"*** 6 ***Ask, "How many hundreds?"*** 3 • Have students complete the subtraction. Record the subtraction including trades on the board as students model them with money manipulatives. ***Say, "What is the difference?"*** \$218	• place-value chart • money manipulatives	23
24	Students who have difficulty adding columns of numbers can look for familiar patterns such as combinations of ten or doubles. • Write "24 + 16 + 17" on the board in vertical form. • Students copy the problem on their paper. ***Ask, "What do we add first?"*** ones place • Direct students to look for tens or doubles. • Model drawing a pointed bracket connecting 4 and 6. Write "10" at the point. ***Say, "Now add 10 + 7."*** 17 • Students write "7" in the ones place and add the ten to the tens place. ***Ask, "How many tens?"*** 5 ***Ask, "What is the sum?"*** 57 • Repeat the procedure with 18 + 34 + 24, and combine doubles first this time.	• pencil and paper	24
25	Divide students into groups of four. Assign each student a number: 1, 2, 3, or 4. • Give each team a different amount of money using bills and coins. Direct students to work together to count the bills and coins in groups according to their value. • Any member of the group should be able to present the total value of their money and explain how they counted to get the total value. • Call out a number between 1 and 4. The student assigned that number must present his or her team's answer.	• money manipulatives • pencil and paper	25

Copyright © by Harcourt Achieve. All rights reserved.

Lesson	Activity	Materials	Reteaching
26	If students have difficulty aligning the digits in money amounts, provide grid paper for them to use. • Show students an example of a money subtraction problem aligned on grid paper. • Allow them to use grid paper whenever needed to align problems. • Another option is for students to turn their regular, lined paper sideways so that the lines on the paper are vertical. $ 4 . 3 2 − $ 2 . 4 8 $	• grid paper	26
27	Divide students into groups of three. Choose a group Leader, Recorder, and Presenter. Give each group a comparing and ordering problem that has four three-digit numbers. • Each student uses money manipulatives to find the order of these amounts. The Leader checks the answers. • The Recorder writes down the group's findings. • The Presenter shares the group's results with the class.	• money manipulatives • pencil and paper	27
28	Some students may have difficulty with regrouping across zeros. Organizing their money manipulatives on a place-value chart will help them see the connection between place value and regrouping. • Write "$300 – $143" on the board. ***Ask, "How much do we start with?"*** $300 • Students model $300 using money manipulatives in the appropriate spaces on their place-value chart. ***Ask, "How much are we taking away?"*** $143 • Students make the trades and place the manipulatives in the appropriate spaces on their place-value chart. • Have students complete the subtraction. Record the subtraction including trades on the board as students model the problem with money manipulatives. ***Ask, "What is the difference?"*** $157	• place-value chart • money manipulatives	28
29	If students do not understand numerators and denominators, they can use the phrase "__ out of__ " when they read a fraction. ***Say, "Look at the rectangle."*** ***Ask, "How many parts are shaded?"*** 1 ***Ask, "Out of how many equal parts?"*** 2 ***Say, "Read the fraction."*** • Prompt students to read "1 out of 2." ***Say, "Look at the circle."*** ***Ask, "How many parts are shaded?"*** 1 ***Ask, "Out of how many equal parts?"*** 3 ***Say, "Read the fraction."*** 1 out of 3	• picture of a rectangle with 1/2 shaded • picture of a circle with 1/3 shaded	29
30	Have students place numbers on a number line with only the tens labeled.	• number line • sticky notes	30

Copyright © by Harcourt Achieve. All rights reserved.

Lesson	Activity	Materials	Reteaching
31	Students will work independently and then in pairs to reinforce writing problems involving directions. • Each student writes a word problem that gives directions from one location to another on the activity map. • Students form pairs and switch problems to answer. • The writer checks the solution and asks the other student to explain how he or she solved the word problem.	• map • pencil and paper	31
32	Give students digit number cards marked from 0 to 9 and one card with a comma printed on it. ***Say, "We are going to practice making numbers in the thousands. I will say a number, and I want you to arrange the digits on your desk to form the number. A comma comes where I say 'thousand' because the comma separates the thousands from the hundreds."*** • After saying each number and giving students sufficient time to form the number, invite a volunteer to write the answer on the board or overhead. ***Say, "Five thousand, seven hundred sixteen."*** 5,716 ***Say, "Twenty-three thousand, six hundred nine."*** 23,609 ***Say, "Five hundred seventy-two thousand, one hundred forty-six."*** 572,146 ***Say, "Thirty-nine thousand, seventeen."*** 39,017	• digit number cards 0–9 • card with comma on it (These can be made with index cards or other cards of similar size.)	32
33	Some students may have difficulty seeing the pattern on a number line when some points are not labeled. Teach students to label all major points that are marked with a dot. • Students copy the number line from the Example, including all numbers, letters, and dots. • Students label each tick mark that is marked with a dot on the number line: 122, 124, 126, 128, 132, 134, 136, 138. ***Ask, "What number is at letter C?"*** 126 ***Ask, "What letter do you see at 134?"*** F • Encourage students to copy number lines and add labels as they complete Lesson Practice items.	• refer to the Example	33
34	Some students may have difficulty visualizing units of measurement. This activity will help them make the connection between nonstandard units of measurement and the units on a standard ruler. • Show students, using a ruler, that a cube is one inch wide. • Demonstrate how to use cubes to measure the paper strip by placing the cubes side by side along the edge. ***Ask, "How long is the paper strip?"*** 6 cubes ***Ask, "How long is that in inches?"*** 6 inches • Measure the strip using a ruler to verify that it is 6 inches long. • Have students measure other objects, first using cubes, and then using a ruler.	• 1-inch cubes • inch ruler • a strip of paper 6 inches long • variety of objects (such as books, glue stick, pencils, markers, crayons)	34

Copyright © by Harcourt Achieve. All rights reserved.

Lesson	Activity	Materials	Reteaching
35	Divide students into groups of four. Choose a Leader, Recorder, Checker, and Presenter for each group. • Each group chooses a small object or picture to measure to the nearest quarter inch. The Leader makes sure that all group members are on task. • The team measures the object and the Recorder writes the measurement on paper. • The Checker verifies that the measurement is correct. • The Presenter shares the results with the class.	• paper • ruler	35
36	If some students require more help with "some and some more" stories, encourage them to write down the information they know. • Read this story: *Terry had $8. She got more money for her birthday. Then she had $15. How much money did Terry get for her birthday?* • Explain to students that many story problems have a beginning, a middle, and an end, just like the stories they read in their reading books. We will draw boxes to represent the beginning, middle, and end of the story problem. ***Ask, "How much money did Terry have at the beginning of the story?"*** $8 ***Say, "We do not know how much money she got in the middle of the story, so we will put a question mark in that space."*** ***Ask, "How much money did she have at the end of the story?"*** $15 • This is a "some and some more" story, so we will write an addition sign and an equal sign. • Have students find the number that goes in the middle box. $7 [$8] + [?] = [$15] Beginning Middle End		36
37	Students who have a difficult time estimating lengths and distances will benefit from more concrete measurement experiences using nonstandard units of measurement. • Students place one paper clip next to the object they are measuring and make an estimate of its length. • They record their estimate of the length of the object using paper clips as the unit of measurement. • Students line up paper clips end to end along the edge of the object to get the exact measurement. • They record that measurement.	• 1-inch cubes • paper clips • small identical objects to use as non-standard units of measurement • variety of objects to measure such as books, glue sticks, pencils	37

Copyright © by Harcourt Achieve. All rights reserved.

Lesson	Activity	Materials	Reteaching
38	Provide students with several digital times written on notecards. Students must then demonstrate the given time on their student clocks.	• notecards with digital times • student clocks	38
39	• Have students make two different bars with connecting cubes (12 and 15). • Students should discuss the difference between the two bars. • Have students write a comparison story problem that involves 12 and 15.	• pencil and paper • connecting cubes	39
40	If some students require more help with subtraction stories, encourage them to write down the information they know in this format. • Read this story: *Marcie's mom gave her some money to spend at the carnival. She spent $8 on food at the carnival. Then she had $15. How much money did Marcie's mom give her?* • Explain to students that many story problems have a beginning, a middle, and an end, just like the stories they read in their reading books. We will draw boxes to represent the beginning, middle, and end of the story problem. ***Ask, "How much money did Marcie have at the beginning of the story?"*** We do not know. ***Say, "We will put a question mark in the first box."*** ***Ask, "How much money did she spend in the middle of the story?"*** $8 ***Ask, "How much money did she have at the end of the story?"*** $15 ***Say, "We will write $15 in the last box."*** • This is a "some went away" story, so we will write a minus sign and an equal sign. • Have students find the number that goes in the first box. 23 ? − $8 = $15 Beginning Middle End		40
41	Divide students into groups of four. Have students choose a partner from the group. Draw two fraction circles on the board and shade each one to model a different fraction. • One partner in each pair names the first fraction, while the other partner names the second fraction. • Each partner interviews the other by asking questions about how to name the fraction. (Example: How many parts are shaded?) If students need help, direct them to use the fraction manipulatives. • After both partners have interviewed each other, they share their answers with the team and compare solutions.	• fraction manipulatives • paper	41
42	Students draw examples and nonexamples of a given fraction. Students then explain why each is an example or a nonexample.	• pencil and paper • crayons	42

Copyright © by Harcourt Achieve. All rights reserved.

Lesson	Activity	Materials	Reteaching
43	If some students require additional support to compare fractions, provide them with this visual aid. • Write "3/5" and "1/2" on the board for students to compare. • Students find the bars that show fourths and halves, and then they trace those rectangles. • They shade one rectangle to show 3/5 and the other to show 1/2. ***Ask, "Which shaded part is larger?"*** 3/5 ***Ask, "How would we write that?"*** 3/5 > 1/2 • Repeat with other fraction pairs. Keep the fraction bars handy for students to trace for other practice problems.	• fraction bars divided into halves, thirds, fourths, fifths	43
44	If students have difficulty understanding the relationship between the numerator and the denominator in a fraction, suggest this alternate way of reading a fraction. • Students draw a circle, divide it into four equal parts, and shade three of the parts. ***Ask, "How many parts are there?"*** 4 ***Ask, "How many parts are shaded?"*** 3 • Tell students that since three out of four parts are shaded, they can read the fraction as "three 'out of' four." • Have students draw circles showing 1/2 and 2/4 and practice reading the fractions by reading the fraction bar as "out of."	• pencil and paper	44
45	If students have difficulty with the concept of probability, have them participate in this hands-on activity. • Show students that you are placing two blue tiles and eight red tiles in the bag. • Have one student take a tile out of the bag without looking. • Have students decide if the color chosen was more likely or less likely to be drawn. Continue until each student has had a turn. • Repeat the activity, but this time do not tell students in advance how many tiles of each color are in the bag. • As each student takes a tile out of the bag, have that student predict what color is more likely or less likely to be drawn. Continue until all the tiles are drawn and have students check their predictions.	• color tiles • paper bags	45
46	Divide students into groups of four. Have students choose a partner from the group. Give each group fraction manipulatives and paper. Write four fractions that are equal to 1 on the board. • One partner in each pair chooses a fraction from the board, models it with fraction manipulatives, and then draws the model and shades it to show that the fraction equals 1. Each partner does the same with a different fraction from the board. • Each partner interviews the other by asking questions about fractions equal to 1. (Examples: How many equal parts are in the fraction? How many parts are shaded?) • After both partners have interviewed each other, they share their answers with the team and compare solutions.	• fraction manipulatives • paper	46

Copyright © by Harcourt Achieve. All rights reserved.

Lesson	Activity	Materials	Reteaching
47	Have students form pairs to discuss finding equivalent fractions. • Guide students through Example 1. Then write a similar problem on the board. Instruct students to think about how to find equivalent fractions. • Students work in pairs to write the answer to the problem. • Volunteers share their solutions with the class.	• paper	47
48	If some students have difficulty understanding the concept of fractions on a number line, have them make a paper number line as follows: • Draw a line from one end of a piece of paper to the other end. • Draw a tick mark on the line close to the left edge and label it "0." • Draw a tick mark on the line close to the right edge and label it "1." • Fold the paper in half. Then unfold it and draw a tick mark on the fold line. Label it "1/2." • Count the line segments (separated by tick marks) between 0 and 1. • Fold the paper in half along the original fold line. Then fold it in half again. Unfold it and draw tick marks on the new fold lines. Label the first tick mark "1/4." Label the second tick mark "3/4." • Count the total number of line segments (separated by tick marks) between the 0 and the 1.	• unlined paper	48
49	If some students require additional support to compare fractions, provide them with this visual aid. • Write "2/5" and "1/2" on the board for students to compare. • Students find the bars that show fifths and halves and trace those rectangles. • Then they shade one rectangle to show 2/5 and the other to show 1/2. ***Ask, "Which shaded part is larger?"*** 1/2 ***Ask, "How would we write that?"*** 1/2 > 2/5 • Repeat with other fraction pairs. Keep the fraction bars handy for students to trace for other practice problems.	• fraction bars divided into halves, thirds, fourths, fifths	49
50	Have students get their money manipulatives and pair off into groups of two. • In each group one student will mix different numbers of $1, $10, and $100 bills. The other student will then pick one bill without looking. • Both students should then investigate their bills to determine how to describe the bill that was chosen. One kind of bill could be more or less likely than the others to be chosen.	• money manipulatives	50

Copyright © by Harcourt Achieve. All rights reserved.

Lesson	Activity	Materials	Reteaching
51	If students have difficulty remembering the attributes of a rectangle, they may be helped by having a chart with the following questions displayed in the classroom: **Is it a Rectangle?** **Is it flat?** **Does it have four sides?** **Does it have four right angles (square corners)?** • Review the attributes of a rectangle listed on the chart. • Examine a variety of two- and three-dimensional figures. • For each figure, ask the questions on the chart to determine if it is a rectangle.	• pre-made chart	51
52	Have students form pairs to practice identifying rectangles. • Draw a rectangle and a polygon that is not a rectangle on the board. Direct students to think about how they could decide if either of the figures is a rectangle. • Students work in pairs to write the answer. Volunteers share their answers with the class.	• paper • ruler	52
53	Pair students before you teach Lesson 53. Then teach the lesson, stopping before you go over Example 2. • Partners discuss strategies to solve the repeated addition problem using the grid paper. • Volunteers share their strategies with the class.	• grid paper	53
54	Some students may have difficulty with the concept of multiplication. This manipulative activity will help them visualize the multiplication pattern. • Write "4 × 7" on the board. • Students build a rectangle using the square tiles to show the multiplication. ***Ask, "How many rows?"*** 4 or 7 ***Ask, "How many columns?"*** 7 or 4 ***Ask, "How many altogether?"*** 28 • Repeat with other multiplication facts.	• square tiles	54
55	Students who have trouble tracing down the columns on the multiplication table in this lesson might benefit from a modified multiplication table with alternate columns highlighted. ***Say, "Find the 1s column. What pattern do you see?"*** count up by 1 ***Say, "Highlight the 1s column."*** ***Say, "Find the 2s column. What pattern do you see?"*** count up by 2 ***Say, "Find the 3s column. What pattern do you see?"*** count up by 3 ***Say, "Highlight the 3s column."*** • Continue as above, highlighting alternate columns. • Allow students to use this modified multiplication table in lesson problems.	• multiplication table, 1 copy per student • highlighters	55

Copyright © by Harcourt Achieve. All rights reserved.

Lesson	Activity	Materials	Reteaching
56	Many students enjoy using flash cards to quiz each other on multiplication facts. Be sure to use only the flash cards with 0s, 1s, or 10s multiplication facts for this lesson. Students can make their own personal flash cards for facts that are difficult for them.	• blank cards • pencils	56
57	Students who have difficulty understanding the concept of multiplication can read the multiplication symbol as "groups of." • Write "3 × 4" on the board. Students copy the fact on their paper. • Guide students to draw a representation of the multiplication. They might draw 3 rows of 4 or 3 circles with 4 tally marks in each circle. • Point out that 3 × 4 is represented as three groups of four. • Complete the multiplication fact on the board (3 × 4 = 12). • Read the fact, replacing the multiplication symbol with the words *groups of.* Touch each number and symbol as you say "3 groups of 4 equals 12." • Repeat with other multiplication facts. Have students read the facts, replacing the multiplication symbol with the words *groups of.*	• pencil and paper	57
58	Some students may have difficulty remembering to add all sides of a rectangle to find the perimeter when only two sides are labeled. Teach them to add labels to all sides before calculating the perimeter. ***Ask, "How do we find the perimeter?"*** add the lengths of all the sides • Draw students' attention to the picture of a rectangle. Point out that only two sides have measurement labels. ***Ask, "What do we know about rectangles that will tell us the length of the other two sides?"*** opposite sides are equal • Add the missing measurement labels to the picture. • Suggest that students add missing labels when calculating perimeters in this lesson.	• pictures of rectangles with two sides labeled (students can copy pictures from the lesson)	58
59	Some students will have difficulty keeping track of the number of 5s when skip counting to multiply. Here is one strategy they might use. • Write "6 × 5" on the board. • Remind students that this is the same as 6 groups of 5. ***Ask, "How can we count 6 groups of 5?"*** skip count by 5 six times • The student holds up 6 fingers and counts a group of 5 for each finger. ***Ask, "What is 6 × 5?"*** 30 • Practice with other "× 5" facts. • Continue to encourage students to memorize all multiplication facts.		59

Copyright © by Harcourt Achieve. All rights reserved.

Lesson	Activity	Materials	Reteaching
60	Some students may have difficulty visualizing a story problem. Drawing a picture of the problem may make it more concrete for them. • Write this story problem on the board: *There were 5 students at Sophie's table. She gave each of them 4 pencils. How many pencils did Sophie give the students at her table?* ***Ask, "What are we counting?"*** pencils ***Ask, "How many groups of pencils?"*** 5 • Students draw a circle for each group. ***Ask, "How many pencils in each group?"*** 4 • Students draw four tally marks in each circle to represent the pencils. ***Ask, "How many pencils did Sophie give the students at her table?"*** 20	• pencil and paper	60
61	Give each student a piece of grid paper. Have them write the squares of the numbers from 1 through 12. • Have students outline the number of boxes on the paper to show the square represented by the multiplication fact. • Suggest that students label each small square within one of the larger squares.	• grid paper	61
62	Some students may need additional real-world experiences to internalize the meaning of area. • Use tiles to build rectangles with 7 rows of 5, 3 rows of 5, and 6 rows of 4. • Find the area of each rectangle by skip counting the rows or columns. • Some students may need to actually move each group of tiles as they count. • Write the area on a card or small paper and place it on top of the tiles.	• color tiles, 35 per student	62
63	After guiding the students through Example 2, write three similar area problems on the board. Divide the students into groups of four. • Each group works together to solve the first problem. • Divide groups into pairs to solve the second problem. • Direct students to work on the last problem independently and then to check their solutions with the group.	• paper	63
64	Provide students with a sheet of paper with the 9s facts, through 9×12, written in column form. The first three facts are shown below: $9 \times 1 = 9$ $9 \times 2 = 18$ $9 \times 3 = 27$ • Have students find as many patterns as possible.	• paper with nines tables	64

Copyright © by Harcourt Achieve. All rights reserved.

Lesson	Activity	Materials	Reteaching
65	Some students may have difficulty with the vocabulary in this lesson. Help them develop a visual reminder. • Students draw a small box in the bottom left corner of the card to indicate a right angle. Label this "right angle." • In the middle of the card, students draw an acute angle and label it. • On the right side of the card, students draw an obtuse angle and label it. • Students can use this card whenever they need to name an angle. They place the "right angle" corner of the card on the angle. They can easily see that if the card fits the angle, it is a right angle. If the angle shows outside the card, it is an obtuse angle. If they can only see one side of the angle, it is an acute angle. acute angle obtuse angle right angle	• index cards, 1 per student	65
66	Some students may have difficulty with the vocabulary in this lesson. Help them develop a visual reminder. • Write "Parallelogram" on the board and have students copy it at the top of their card. • Students draw a rectangle. They trace each pair of opposite sides with a crayon, using a different color for each pair. For example, they might trace the top and bottom sides with a red crayon, and the right and left sides with a blue crayon. • Students draw a square and trace each pair of opposite sides with a crayon, using a different color for each pair. • Students draw a parallelogram with two obtuse and two acute angles, tracing opposite sides as above. • Students can refer to this card whenever they need to identify a parallelogram.	• index cards • crayons	66
67	Some students may have difficulty remembering the names of common polygons. Students can work in pairs to practice. • Each student should create a set of flash cards with the name of the polygon on one side, and a picture of the corresponding shape on the other. • Place students in pairs and have them take turns identifying the names of each of the polygons on the flash cards. Encourage students to help each other find ways to remember the names of each shape.	• index cards	67
68	Students who have difficulty identifying congruent shapes will benefit from this hands-on activity. • Place students in groups and give them construction paper. Have each student cut out pairs of identical shapes. These will represent congruent figures. • Have each group mix together their pairs of shapes and spread them out. Then have students take turns finding congruent pairs from the pile.	• construction paper • scissors	68

Copyright © by Harcourt Achieve. All rights reserved.

Lesson	Activity	Materials	Reteaching
69	Some students may have difficulty classifying the different kinds of triangles. Making flash cards and practicing with them will give students extra practice with this skill. • Place students in pairs and hand them index cards. Have students write the type or name of a triangle on one side and draw the shape on the other side. • Have students take turns identifying the name or feature that matches each triangle on the card.	• index cards	69
70	Divide students into groups of four. • Mix the cards up, and place them in a stack face down. • Each player takes turns flipping a card over. The player with the largest product keeps all four cards. If two players have the same high product, those players each flip another card. The highest product wins all cards. • When all cards in the middle stack have been used, the player with the most cards wins.	• paper • flash cards from Lesson Activity 25	70
71	Some students may have difficulty identifying the attributes of a three-dimensional shape in a picture. These students will benefit from more experiences that connect tactile examination of the object with visual examination of the picture. • Students match the object to the picture. • Students identify and count the faces, edges, and vertices on the object. • Then students identify and count the faces, edges, and vertices on the picture.	• rectangular prism • cube • pictures of a rectangular prism and a cube	71
72	Provide students with 36 cubes and have them build as many different stacks as possible with all 36 cubes.	• cubes, 36 per student	72
73	Some students may need more concrete experiences with finding volume. • Students cover the bottom of a box with cubes in rows and columns. • They repeat that layer until they get to the top. • Students remove the cubes from the box and reassemble the stack so that they can see all the blocks. • They count the cubes in the top layer and skip count or add to find the total number of cubes in all the layers.	• variety of small boxes that can be filled with cubes • 1-inch cubes	73
74	Divide students into groups of four. Choose a Leader, a Recorder, a Checker, and a Presenter. • After guiding students through Example 2, direct groups to make a list of different objects that weigh about one ounce, one pound, and one ton. There should be at least one object for each unit. • The Leader makes sure all members are on task. The Recorder writes the list. The Checker verifies the weight of the objects either from a book or the Internet. • The Presenter will share the group's list with the class.	• paper • encyclopedia • Internet access	74

Copyright © by Harcourt Achieve. All rights reserved.

Lesson	Activity	Materials	Reteaching
75	Some students may have difficulty visualizing geometric solids by looking at pictures. These students need more tactile experience with the shapes. • Each student selects one geometric solid. • Students trace the flat surfaces of the solid. • Students exchange papers and try to guess which solid their partner traced.	• geometric solids • pencil and paper	75
76	Students will work independently and then in pairs to reinforce multiplication facts. • Direct students to choose four rows of the multiplication table and to make flash cards showing the multiplication problem on one side and the answer on the other. • Have students form pairs, combine and mix their cards, and use the table to solve the problems.	• index cards	76
77	Have students form pairs. Distribute one small rectangular prism and one ruler to each pair. • Instruct students to identify the object's length, width, and height. • Have the students use the ruler to measure the object's dimensions. • Have one of the students record the measurements. • Ask each pair to write a number sentence demonstrating how to calculate the object's volume. ***Ask, "Since the length, width, and height were measured in inches, what units will the volume have?"*** cubic inches	• a variety of small boxes that can be filled with 1-inch cubes • rulers	77
78	Some students may have difficulty multiplying by multiples of 10. Use this hands-on activity with base ten blocks as a concrete representation of this concept. • Write "3 × 40" on the board. • Tell students to make 1 group of 40 using ten-sticks. ***Ask, "How many tens are equal to 40?"*** 4 • Students make 3 groups of 40 using ten-sticks. ***Ask, "How many tens are there altogether?"*** 12 ***Ask, "What is the value of 12 tens?"*** Count by 10 to 120 or regroup to find 1 hundred and 2 tens. • Write the answer to the number sentence 3 × 40 on the board.	• ten-sticks, 20 per pair of students	78
79	Some students may need additional hands-on experience with measurement. Provide opportunities to measure and reinforce the concept of perimeter with this activity. • Students measure each side of the rectangle in centimeters and record the measurement. • Combine the lengths of the sides and record the perimeter. • Continue as time allows.	• ruler • rectangles of varying sizes	79

Copyright © by Harcourt Achieve. All rights reserved.

Lesson	Activity	Materials	Reteaching
80	Some students may need more concrete experience with units of mass. • Remind students that one large paper clip has a mass of about one gram. • Students place one of the small objects on one side of the balance scale. • Students place enough large paper clips on the other side of the scale to reach a balance. • Students count the number of large paper clips to determine the approximate mass of the object in grams.	• balance scale • several large paper clips • small classroom objects such as pencil, marker, scissors, glue stick	80
81	Students who are still struggling with multiplication facts can use a multiplication table for lessons involving multi-digit multiplication. This will give them an opportunity to learn the algorithm without being hindered by not knowing the basic facts. • Write these multiplication problems on the board: 24×2 43×2 29×2 38×2 • Students multiply, starting in the ones place. • They use the multiplication table for any facts they do not know. • If regrouping is necessary in the ones place, students write +1 above the digit in the tens place to remind them to add the ten. • Students check their work with repeated addition. $24 \times 2 = 48$ $24 + 24 = 48$	• multiplication table	81
82	Students who have difficulty dividing can use this strategy. • Students copy this division problem: $24 \div 2$ $2\overline{)24}$ • They draw two circles on their paper. • Then they "deal" 24 tally marks by drawing one in the first circle, then drawing one in the second circle, and so on, alternating circles, counting as they go. • When students have drawn 24 tally marks, they count the marks in one circle. That number is the quotient.	• pencil and paper	82
83	Some students might have difficulty isolating numbers on the 2s row of the multiplication table. Give them a visual aid while reviewing patterns and skip counting. • Ask what pattern they see on the 2s row. counting up by 2s • Students highlight the 2s row. • Students count by 2s to 24 while touching the numbers on the table.	• multiplication table • highlighter	83

Copyright © by Harcourt Achieve. All rights reserved.

Lesson	Activity	Materials	Reteaching
84	Some students may have difficulty multiplying two-digit numbers. Use money manipulatives to help them with regrouping in multiplication. • Write "27 × 4" in vertical form on the board. ***Say, "Another way to say this is '4 groups of 27.'"*** • Students make 4 groups of $27. • Students combine the $1 bills first and regroup ten $1 bills as a $10 bill. • Show how the regrouping is recorded in the example on the board. • Students combine the $10 bills. Record the answer on the board.	• money manipulatives ($1 bills and $10 bills)	84
85	Some students may have difficulty visualizing division situations. This activity will give them additional hands-on experience. • Tell students a simple division story, such as, *I have 12 pencils. I want to give 6 students an equal number of pencils. How many pencils will each student get?* ***Ask, "How many in all?"*** 12 • Students count out 12 color tiles. ***Ask, "How many groups?"*** 6 • Students put one index card per group on their desk. • Students separate 12 color tiles equally among the six index cards.	• color tiles, 25 per pair of students • index cards, 6 per pair of students	85
86	Some students may have difficulty remembering which symbols to use when recording fact families. Provide them with this visual aid. • Write the following template on the board. Have students copy it on their index card. **Fact Family** ____ × ____ = ____ ____ × ____ = ____ ____ ÷ ____ = ____ ____ ÷ ____ = ____ • Students use the template to write a fact family for 3, 7, 21. • Find the product of 4 × 8, and then write related multiplication and division facts using the template.	• index cards, student	86

Copyright © by Harcourt Achieve. All rights reserved.

Lesson	Activity	Materials	Reteaching
87	Have students create a "Gallon Guy" and label the parts. Mr. Gallon is a visual aid used to learn and understand Customary capacity units. Mr. Gallon is put together like a human body. Each portion of his body represents a measure, except for his head. The torso represents the gallon, which is the biggest part of the body and unit of capacity to which everything is connected. Connected directly to the gallon are four quarts representing arms and legs. This shows that four quarts equal a gallon. In a human body, the bottom portions of arms and legs have two separate bones, and so does Mr. Gallon. Two pints are connected to each quart. Therefore, there are two pints in each quart and eight pints in a gallon. The last parts of Mr. Gallon are the fingers and toes that represent cups. The fingers are the smallest portions on the body, as cups are the smallest unit. Two cups are attached to each pint. Therefore, there are 16 cups in a gallon, 4 cups in a quart, and 2 cups in a pint.	• construction paper • scissors • glue or tape • pencil	87
88	Provide students with a hundred-number chart and have them color all the even numbers blue and all the odd numbers red.	• hundred-number chart • crayons (blue and red)	88
89	Students can use their rulers to divide numbers less than 30. • Have students place their rulers on a sheet of paper so that they are measuring in centimeters. They should draw a line 30 cm long, marking the points at 0 cm and at 30 cm. • To divide 30 into equal groups of 6, have students count 6 cm from the left and make a mark on their line. From the 6-cm mark, have students count another 6 cm, make a mark, and so on. • When they are finished, students can count the number of segments on their line. The answer can be checked using a multiplication table.	• rulers • pencil and paper • multiplica-tion table	89
90	Some students may have difficulty visualizing a story problem. Drawing a picture of the problem may make it more concrete for them. • Write this story problem on the board: *Rachel had 24 erasers. She wanted to divide them equally among the 6 other students at her table. How many erasers would each student get?* ***Ask, "What are we counting?"*** erasers ***Ask, "How many groups of erasers?"*** 6 • Students draw a circle for each group. ***Ask, "How many erasers are there altogether?"*** 24 ***Ask, "What will we do with the 24 erasers?"*** divide them equally among 6 students • Students draw a tally mark in each circle, counting as they add one more to each circle until they have drawn 24 tally marks. ***Ask, "How many erasers did each student get?"*** 4	• pencil and paper	90

Copyright © by Harcourt Achieve. All rights reserved.

Lesson	Activity	Materials	Reteaching
91	Some students may have difficulty visualizing the relationship between a partially filled jar and a full jar. This activity will give them an opportunity to solve an easier estimation problem. • Give each pair of students one box and a set of counting cubes. Ask students to fill the bottom of the rectangular container with counting cubes. • Students count the number of cubes they used to fill the bottom of the box. • Students stack one column of cubes to determine how many layers will fill the box. ***Ask, "How many layers will fill the box?"*** See student work. ***Ask, "What fraction of the box is filled with the first layer?"*** See student work. • Students add or multiply to estimate the number of blocks that will fill the box.	• 1-inch counting cubes, 20 per pair of students • rectangular prism such as a small tissue box or supply box	91
92	Some students may have difficulty recognizing compatible numbers. This activity will help them learn to pair numbers with sums that end in zero. • Give each student a set of number cards. Have students lay the cards out in random order. Instruct students to choose sets of two cards whose sum equals 20. • Students should write each addition fact represented by their cards on a sheet of paper. • If this activity proves too difficult, use number cards with digits 0–10 and find sums of 10.	• number cards 0–20, 1 set per pair of students	92
93	Some students may have difficulty rounding three-digit numbers. This activity pairs tactile and visual learning to teach and reinforce rounding. • Give students a set of number cards. Write numbers on the board that do not have repetitive digits, such as 264, 709, and 58. Have students model the numbers using their number cards. • Instruct students to point to the second digit of the number. ***Ask, "Is this digit 5 or larger?"*** • If so, have the student replace the first digit with the next-higher number card. Have them replace all the other digits with the zero cards. • Have students repeat the process using different sets of numbers: 123, 456, and 789; 147, 256, and 369; 159, 267, and 348.	• number cards 0–9, plus two extra zero cards	93
94	Some students may have difficulty remembering the multiples of 25. Have them make a visual reminder and keep it handy to help with problems involving compatible numbers. • List the multiples of 25 on the board for students to copy and use as a reference. • Have students find the multiples of 25 closest to 21, 69, 121, and 98. • Let students use their list of multiples of 25 to help them with Lesson Practice and Written Practice.	• pencil and paper	94

Copyright © by Harcourt Achieve. All rights reserved.

Lesson	Activity	Materials	Reteaching
95	If some students have difficulty understanding the word *reasonable,* connect it to concrete concepts. Read the statements below. Have students indicate if the statements are reasonable (thumbs up) or not reasonable (thumbs down). • A small car weighs about 1 ton. • A new bicycle costs 35¢. • A family can eat a dozen apples in a week. • The temperature outside is 94° and it is snowing. • There are 128 people in our classroom.		95
96	Divide students into groups of four. Assign each student a number: 1, 2, 3, or 4. • Guide students through Example 1. Then write a similar problem on the board. Direct students to work together to solve the problem. • Any member of the group should be able to present the answer to the problem and explain how they solved it. • Call out numbers between 1 and 4. Each member assigned the number you call out must present his or her group's solution and strategy for solving the problem.	• pencil and paper	96
97	Some students may be confused if regrouping is necessary in both the tens and hundreds places when multiplying a three-digit number by a one-digit number. Use this template as a visual reminder. • Write the template below on the board and have students copy it on their card. ( )( ) _ _ _ × _ _, _ _ _ • •Demonstrate using the template below to solve 263 × 7. (4) (2) 2 6 3 × 7 1, 8 4 1 • Remind students to use the template to solve similar problems until they can keep track of the numbers on their own.	• index cards, 1 per student	97

Copyright © by Harcourt Achieve. All rights reserved.

Lesson	Activity	Materials	Reteaching
98	Some students may have difficulty organizing information. Provide these students with a table template to write on. Keep a collection of templates on hand for students to use when needed. • See Example 1. • Students write "Number of Pennies" in the first row and "Mass of Pennies" in the second row. • Write "400" in the second column of the first row. • Skip count by 400 to complete the first row. • Complete the second row with these masses: 1 kg, 2 kg, 3 kg, 4 kg, 5 kg. Number of Pennies: 400 | 800 | 1,200 | 1,600 | 2,000 Mass of Pennies: 1 kg | 2 kg | 3 kg | 4 kg | 5 kg	• blank two-row table	98
99	Have students use their money manipulatives to help them visualize the effects of estimation. • Instruct students to solve the problems below using their manipulatives. They should first round the values to the nearest dollar amount and determine an estimate. Then students should find the exact answer and compare it to the estimate. $2.65 + $4.99 = $7.64; $3 + $5 = $8; estimate is greater by 36¢ $3.10 × 3 = $9.30; $3 × 3 = $9; estimate is less by 30¢ $6.58 × 4 = $26.32; $7 × 4 = $28; estimate is greater by $1.68	• money manipulatives	99
100	Students will work independently and then in pairs to reinforce multiplying with money. • Guide students through Example 2. • Then instruct them to write their own word problem using money and multiplication. • Have students form pairs and solve each other's multiplication problem. • Students should explain their thinking to their partner.	• pencil and paper	100

Copyright © by Harcourt Achieve. All rights reserved.

Lesson	Activity	Materials	Reteaching
101	Some students may have difficulty remembering the sequence of steps for the division algorithm. Have them make a card with this mnemonic: Dad, Mother, Sister, Brother. **D**ad - **D**ivide **M**om - **M**ultiply **S**ister - **S**ubtract **B**rother - **B**ring down • Write the mnemonic above on the board and have students copy it on their cards. • Write this division problem on the board and practice using the steps: $5\overline{)70}$ **Step 1:** Divide 7 by 5 $5\overline{)70}$ → 1 **Step 2:** Multiply 1 × 5 $5\overline{)70}$ → 1, 5 **Step 3:** Subtract 5 from 7 $5\overline{)70}$ → 1, −5, 2 **Step 4:** Bring down the 0 $5\overline{)70}$ → 1, −5, 20	• index cards, 1 per student	101
102	Some students may need more experience with sorting and classifying. Guide them through this hands-on activity. • Instruct students to sort their tiles into two groups based on a rule. • Have students trade places with a partner and identify each other's sorting rule.	• pattern blocks, 20 per student	102
103	Have students individually answer Example 2 without looking at the answers on the next page. Then divide the students into groups of four. • Students take turns sharing their strategies with the group. All strategies are acceptable. • One person from the group should be chosen as the Recorder. The group works together to write a summary of their strategies.	• pencil and paper	103
104	Some students may have difficulty naming geometric figures in pictures and identifying their attributes. This sorting activity will give them an opportunity to practice both skills. • Students work with a partner to name the figure on each flash card. They can look at the back of each card to self-check. • After each group has named the shapes, they sort them and describe their sorting rule.	• pre-made flash cards with pictures of geometric figures on one side and names of the figures on the other	104

Copyright © by Harcourt Achieve. All rights reserved.

Lesson	Activity	Materials	Reteaching
105	If students have difficulty understanding the concept of classification using a Venn diagram, provide them with these templates as a concrete visual model. • Provide each student with one worksheet. Have them begin by working through each of the three examples in the lesson using the appropriate sorting diagram. • Have the students attempt the two practice problems with the remaining diagrams. • Instruct the students to exchange the last two diagrams with another student and to check their work. For any mistakes that are made, have the two students pair up to solve the problem correctly. *Note*: For the first few examples on the worksheet, fill in the labels and one or two numbers inside the circles. For the practice problems, provide blanks for students to fill in the labels. Be sure the overlapping parts of the circles have enough room for students to fill in the number or picture.	• worksheet with sorting circles and Venn diagrams for the examples and practice problems in this lesson (see *Note*)	105
106	Some students may have difficulty processing the abstract concept in this lesson. Use this hands-on activity with these students. • Using a figure from this lesson, have one student place square tiles on all of the whole squares inside the figure while another student keeps count. Then repeat this process with the half squares. • Have the group add up their total, making sure to correctly add the half squares. • For additional practice, draw other irregular figures on grids for students to use. *Note*: For tiles, make about 25 whole squares and 10 half squares, cut along a diagonal, for each group of students.	• worksheets with reproductions of the figures on grids from this lesson • tiles made by cutting up a blank grid on colored paper	106
107	Some students may have difficulty drawing. • Have students use the simpler straight-line figures for the enlargement activity rather than pictures with curves. *Note*: Use figures like those in Lesson 106, Example 1 or problem **a** in Lesson Practice.	• handouts with simple straight-line figures drawn on small grids	107
108	Geoboards can be used to form shapes for this activity. Students should start with relatively easy shapes like triangles or polygons, and then create more challenging, irregular shapes.	• geoboards • rubber bands	108

Copyright © by Harcourt Achieve. All rights reserved.

Lesson	Activity	Materials	Reteaching
109	Using a real-world, concrete example of grids and coordinates should help some students who are having difficulty with the more abstract concept of points on a plane. • Find an easily identifiable city, street, or place on the map and have students identify the grid coordinates (for example, A-2, C-3, and so on) of the region it is in. • Use the street locator on the map to have students find particular streets by locating the street's region from its coordinates. • If maps still prove too difficult for some students, switch to Bingo cards. Call out a space on the card and have the students locate the space and write down its numeric coordinates. Do this for seven or eight different spaces.	• maps that have letters and numbers labeling the horizontal and vertical regions of a grid (like those from a phone book) or bingo cards	109
110	Divide students into groups of four. Choose a group Leader, Recorder, Checker, and Presenter. Give each group a blank sheet of grid paper. Direct groups to plot points on the grid to make their own dot-to-dot design. • The Leader makes sure that all group members are on task. • The Recorder works with the group to record the coordinates for the points on the grid. • The Checker verifies the coordinates to make sure the line segments form the intended picture. • The Presenter shares the group's new dot-to-dot activity and asks a volunteer from a different group to solve it.	• grid paper	110

Copyright © by Harcourt Achieve. All rights reserved.